18th Edition
IET Wiring Regulations
Explained and Illustrated

This popular guide focuses on common misconceptions in the application of the IET Wiring Regulations. It explains in clear language those parts of the regulations that most need simplifying, outlining the correct procedures to follow and those to avoid. Emphasis has been placed on areas where confusion and misinterpretation are common, such as earthing and bonding, circuit design and protection, and in particular the increased use of RCDs.

With the content covering the requirements of both City & Guilds and EAL courses and containing sample exam questions and answers, this book is also an ideal revision guide.

Brian Scaddan, I Eng, MIET, is an Honorary Member of City & Guilds and has over 45 years' experience in further education and training. He was Director of Brian Scaddan Associates Ltd, an approved training centre offering courses on all aspects of electrical installation contracting, including those for City & Guilds and EAL. He is also a leading author of books on other installation topics.

By the same author

18th Edition IET Wiring Regulations: Design and Verification of Electrical Installations, 9th ed, 978-1-138-60600-5

18th Edition IET Wiring Regulations: Electric Wiring for Domestic Installers, 16th ed, 978-1-138-60602-9

18th Edition IET Wiring Regulations: Inspection, Testing and Certification, 9th ed, 978-1-138-60607-4

18th Edition IET Wiring Regulations: Wiring Systems and Fault Finding for Installation Electricians, 7th ed, 978-1-138-60609-8

Electrical Installation Work, 8th ed, 978-1-138-84927-3

PAT: Portable Appliance Testing, 4th ed, 978-1-138-84929-7

The Dictionary of Electrical Installation Work, 978-0-08-096937-4

18th Edition IET Wiring Regulations Explained and Illustrated

11th Edition

Brian Scaddan

LONDON AND NEW YORK

Self-employed

2. A self-employed person is an individual who works for gain or reward otherwise than under a contract of employment whether or not he or she employs others.

Employee

3. Regulation 3(2)(a) reiterates the duty placed on employees by Section 7(b) of the HSW Act.
4. Regulation 3(2)(b) places duties on employees equivalent to those placed on employers and self-employed persons where these are matters within their control. This will include those trainees who will be considered as employees under the Regulations described in paragraph 1.
5. This arrangement recognizes the level of responsibility which many employees in the electrical trades and professions are expected to take on as part of their job. The 'control' which they exercise over the electrical safety in any particular circumstances will determine to what extent they hold responsibilities under the Regulations to ensure that the Regulations are complied with.
6. A person may find him or herself responsible for causing danger to arise elsewhere in an electrical system, at a point beyond their own installation. This situation may arise, for example, due to unauthorized or unscheduled back feeding from the installation onto the system, or to raising the fault power level on the system above rated and agreed maximum levels due to connecting extra generation capacity, etc. Because such circumstances are 'within his or her control', the effect of Regulation 3 is to bring responsibilities for compliance with the rest of the regulations to that person, thus making him or her a duty holder.

Absolute/reasonably practicable

7. Duties in some of the Regulations are subject to the qualifying term 'reasonably practicable'. Where qualifying terms are absent the requirement in the Regulation is said to be absolute. The meaning

of reasonably practicable has been well established in law. The interpretations below are given only as a guide to duty holders.

Absolute

8. If the requirement in a Regulation is 'absolute', for example if the requirement is not qualified by the words 'so far as is reasonably practicable', the requirement must be met regardless of cost or any other consideration. Certain of the regulations making such absolute requirements are subject to the Defence provision of Regulation 29.

Reasonably practicable

9. Someone who is required to do something 'so far as is reasonably practicable' must assess, on the one hand, the magnitude of the risks of a particular work activity or environment and, on the other hand, the costs in terms of the physical difficulty, time, trouble and expense which would be involved in taking steps to eliminate or minimize those risks. If, for example, the risks to health and safety of a particular work process are very low, and the cost or technical difficulties of taking certain steps to prevent those risks are very high, it might not be reasonably practicable to take those steps. The greater the degree of risk, the less weight that can be given to the cost of measures needed to prevent that risk.
10. In the context of the Regulations, where the risk is very often that of death, for example from electrocution, and where the nature of the precautions which can be taken are so often very simple and cheap, e.g. insulation, the level of duty to prevent that danger approaches that of an absolute duty.
11. The comparison does not include the financial standing of the duty holder. Furthermore, where someone is prosecuted for failing to comply with a duty 'so far as is reasonably practicable', it would be for the accused to show the court that it was not reasonably practicable for him or her to do more than he or she had in fact done to comply with the duty (Section 40 of the HSW Act).

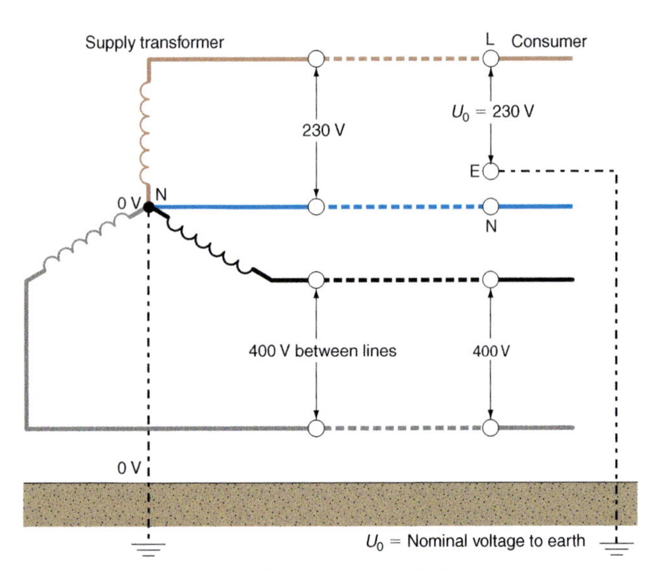

FIGURE 1.1 DNO supply voltages.

Questions

1. Which Part of BS 7671 forms its basis?
2. Which three sets of legislation are most likely to be associated with BS 7671?
3. Under the EAWR 1989, which persons have a duty of care placed upon them?
4. Which item of legislation is solely applicable to electrical installations in dwellings?
5. When may an unqualified installer issue an Electrical Installation Certificate?
6. Under Part 'P' of the Building Regulations, to whom should work in a bathroom be notified?
7. Where, in BS 7671:2008, are Statutory Documents listed?
8. What are the DNO's standard voltages and tolerances?
9. What is the voltage Band and maximum a.c. value of Extra Low Voltage?
10. What is the ESQCR?

Answers

1. Part 1
2. The H&S at Work Act 1974, the EAWR 1989, and Part'P' of the Building Regulations
3. Employers, self employed, employees and managers of mines and quarries
4. Approved Document P of the Building Regulations
5. Never
6. The LABC
7. Appendix 2
8. 230V, 400V; +10%, −6%
9. Band 1; 50V a.c.
10. The Electricity Safety, Quality and Continuity Regulations 2002. These are applicable to the DNOs

CHAPTER 2

Earthing

☞ Relevant BS 7671 chapters and parts: Chapters 31, 41, 54, Part 7

Important terms/topics covered in this chapter:

- **Basic protection** Protection against electric shock under fault-free conditions
- **Bonding conductor** A protective conductor providing equipotential bonding
- **Circuit protective conductor (cpc)** A protective conductor connecting exposed conductive parts of equipment to the main earthing terminal
- **Earth** The conductive mass of earth, whose electric potential at any point is conventionally taken as zero
- **Earth electrode resistance** The resistance of an earth electrode to earth
- **Earth fault current** An overcurrent resulting from a fault of negligible impedance between a line conductor and an exposed conductive part or a protective conductor
- **Earth fault loop impedance** The impedance of the phase-to-earth loop path starting and ending at the point of fault
- **Earthing conductor** A protective conductor connecting a main earthing terminal of an installation to an earth electrode or other means of earthing
- **Equipotential bonding** Electrical connection maintaining various exposed conductive parts and extraneous conductive parts at a substantially equal potential
- **Exposed conductive part** A conductive part of equipment which can be touched and which is not a live part but which may become live under fault conditions
- **Extraneous conductive part** A conductive part liable to introduce a potential, generally earth potential, and not forming part of the electrical installation

18th Edition IET Wiring Regulations: Explained and Illustrated. 978-1-138-60606-7.

EARTHING IN THE IET REGULATIONS (IET REGULATIONS CHAPTER 4, SECTION 411)

In the preceding pages we have briefly discussed the reasons for, and the importance and methods of, earthing. Let us now examine the subject in relation to the IET Regulations.

Contact with metalwork made live by a fault is clearly undesirable. One popular method of providing some measure of protection against the effects of such contact is by protective earthing, protective equipotential bonding and automatic disconnection in the event of a fault. This entails the connection to earth of:

1. All metalwork associated with electrical systems that are not live parts but which could become live under fault conditions. These are termed exposed conductive parts. This is Earthing.
2. The joining together and connection to earth of all metalwork liable to introduce a potential including earth potential, termed extraneous conductive parts. Examples are gas, oil and water pipes, structural steelwork, radiators and baths. This is Bonding.

The conductors used in such connections are called *protective conductors*, and they can be further subdivided into:

1. Circuit protective conductors, for connecting exposed conductive parts to the main earthing terminal.
2. Main protective bonding conductors, for bonding together main incoming services, structural steelwork, etc.
3. Supplementary bonding conductors for bonding exposed conductive parts and extraneous conductive parts, when circuit disconnection times cannot be met, or in special locations, such as bathrooms, swimming pools, etc.

The effect of all this earthing and bonding is to create a zone in which all metalwork of different services and systems will, even under fault conditions, be at a substantially equal potential. If, added to this, there is a low-resistance earth return path, the protection should operate fast enough to prevent danger (IET Regulations 411.3–411.6).

The resistance of such an earth return path will depend upon the system (see the next section), either TT, TN-S or TN-C-S (IT systems will not be discussed here, as they are extremely rare and unlikely to be encountered by the average contractor).

EARTHING SYSTEMS (IET REGULATIONS DEFINITIONS (SYSTEMS))

These have been designated in the IET Regulations using the letters T, N, C and S. These letters stand for:

T	terre (French for earth) and meaning a direct connection to earth
N	neutral
C	combined
S	separate

When these letters are grouped they form the classification of a type of system. The first letter in such a classification denotes how the supply source is earthed. The second denotes how the metalwork of an installation is earthed. The third and fourth indicate the functions of neutral and protective conductors. Hence:

1. A TT system has a direct connection of the supply source neutral to earth and a direct connection of the installation metalwork to earth. An example is an overhead line supply with earth electrodes, and the mass of earth as a return path (Figure 2.7).
2. A TN-S system has the supply source neutral directly connected to earth, the installation metalwork connected to the earthed neutral of the supply source via the lead sheath of the supply cable, and the neutral and protective conductors throughout the whole system performing separate functions (Figure 2.8).
3. A TN-C-S system is as the TN-S but the supply cable sheath is also the neutral, i.e. it forms a combined earth/neutral conductor known as a PEN (protective earthed neutral) conductor (Figure 2.9).

The installation earth and neutral are separate conductors. The DNO's part of this system is known as PME.

Note that only single-phase systems have been shown, for simplicity.

Hence $Z_s = Z_e + (R_1 + R_2)_1 = 0.2 + 0.035 = 0.235\,\Omega$, which is less than the Z_s maximum of 0.24 Ω quoted for a 160 A BS 88 fuse in Table 41.3 of the Regulations.

Motor circuit $(R_1 + R_2)_2$

Here we have 25 m of 25 mm² line conductor with 25 m of 2.5 mm² cpc. Hence:

$$(R_1 + R_2)_2 = \frac{25 \times (0.727 + 7.41) \times 1.2}{1000} = 0.24\,\Omega$$

$$\therefore \text{Total } Z_s = Z_e + (R_1 + R_2)_1 + (R_1 + R_2)_2 = 0.2 + 0.035 + 0.24 = 0.48\,\Omega$$

which is less than the Z_s maximum of 0.91 Ω quoted for a 45 A BS 88-3 fuse from Table 41.3 of the Regulations. Hence we have achieved compliance with the shock-risk constraints.

RESIDUAL CURRENT DEVICES

Additional protection can only be provided by RCDs with a rating $I_{\Delta n}$ not exceeding 30 mA.

The following list indicates the ratings and uses of RCDs detailed in BS 7671.

Requirements for RCD protection

30 mA

- All socket outlets rated at not more than 32 A
- Mobile equipment rated at not more than 32 A for use outdoors
- All domestic circuits supplying luminaires
- All circuits in a bath/shower room
- Preferred for all circuits in a TT system
- All cables installed less than 50 mm from the surface of a wall or partition unless they are adequately mechanically protected or have earthed screening or armouring, and also at any depth if the construction of the wall or partition includes metallic parts
- In zones 0, 1 and 2 of swimming pool locations
- All circuits in a location containing saunas, etc.
- Socket outlet final circuits not exceeding 32 A in agricultural locations

- Circuits supplying Class II equipment in restrictive conductive locations
- Each socket outlet in caravan parks and marinas and final circuit for houseboats
- All socket outlet circuits rated not more than 32 A for show stands, etc.
- All socket outlet circuits rated not more than 32 A for construction sites (where reduced low voltage, etc. is not used)
- All socket outlets supplying equipment outside mobile or transportable units
- All circuits in caravans
- All circuits in circuses, etc.
- A circuit supplying Class II heating equipment for floor and ceiling heating systems.
- Socket outlets up to 63A for onshore units for inland navigation vessels

100 mA

- Socket outlets of rating exceeding 32 A in agricultural locations.

300 mA

- At the origin of a temporary supply to circuses, etc.
- Where there is a risk of fire due to storage of combustible materials
- All circuits (except socket outlets) in agricultural locations.
- Socket outlets over 63A for onshore units for inland navigation vessels

500 mA

- Any circuit supplying one or more socket outlets of rating exceeding 32 A, on a construction site.

We have seen the importance of the total earth loop impedance Z_s in the reduction of shock risk.

However, in some systems and especially TT, where the maximum values of Z_s given in Tables 41.2–41.4 of the Regulations may be hard to satisfy, an RCD may be used: its residual rating being determined from:

$$I_{\Delta n} \leq \frac{50}{Z_s}$$

switch. Better still, especially in TT systems, is the use of a 100 mA RCD for protecting circuits other than socket outlets.

Modern developments in CB, RCD and consumer unit design now make it easy to protect any individual circuit with a combined CB/RCD (RCBO), making the use of split-load boards unnecessary.

Except for domestic premises, additional protection by a 30 mA or less RCD may be omitted if subject of a documented risk assessment.

Supplementary bonding (IET Regulations Section 415.2)

In general the only Supplementary bonding required is for special locations such as bathrooms (not always needed – see Chapter 7), swimming pools, agricultural premises, etc. and where disconnection times cannot be met.

By now we should know why bonding is necessary; the next question, however, is to what extent bonding should be carried out. This is perhaps answered best by means of question and answer examples:

1. **Do I need to bond the hot and cold taps and a metal kitchen sink together? Surely they are all joined anyway?**
   Provided that main protective bonding conductors have been correctly installed there is no specific requirement in BS 7671 to do this.
2. **Do I have to bond radiators in a premises to, say, metal-clad switches or socket outlets etc.?**
   Supplementary bonding is only necessary when extraneous conductive parts are simultaneously accessible with exposed conductive parts and the disconnection time for the circuit concerned cannot be achieved. In these circumstances the bonding conductor should have a resistance $R \leq 50/I_a$, where I_a is the operating current of the protection.
3. **Do I need to bond metal window frames?**
   In general, no. Apart from the fact that most window frames will not introduce a potential from anywhere, the part of the window most likely to be touched is the opening portion, to which it would not

be practicable to bond. There may be a case for the bonding if the frames were fortuitously touching structural steel work.

4. **What about bonding in bathrooms?**
   Refer to Chapter 7.
5. **What size of bonding conductors should I use?**
   Main protective bonding conductors should be not less than half the size of the main earthing conductor, subject to a minimum of 6.0 mm^2 or, where PME (TN-C-S) conditions are present, a minimum of 10.0 mm^2. For example, most new domestic installations now have a 16.0 mm^2 earthing conductor, so all main bonding will be in 10.0 mm^2. Supplementary bonding conductors are subject to a minimum of 2.5 mm^2 if mechanically protected or 4.0 mm^2 if not. However, if these bonding conductors are connected to exposed conductive parts, they must be the same size as the cpc connected to the exposed conductive part, once again subject to the minimum sizes mentioned. It is sometimes difficult to protect a bonding conductor mechanically throughout its length, and especially at terminations, so it is perhaps better to use 4.0 mm^2 as the minimum size.
6. **Do I have to bond free-standing metal cabinets, screens, workbenches, etc.?**
   No. These items will not introduce a potential into the equipotential zone from outside, and cannot therefore be regarded as extraneous conductive parts.

The Faraday cage

In one of his many experiments, Michael Faraday (1791–1867) placed an assistant in an open-sided cube which was then covered in a conducting material and insulated from the floor. When this cage arrangement was charged to a high voltage, the assistant found that he could move freely within it, touching any of the sides, with no adverse effects. Faraday had, in fact, created an equipotential zone, and of course in a correctly bonded installation, we live and/or work in Faraday cages!

Questions

1. What is the conventional potential of the earth?
2. What is the most common method of connection to earth?
3. Which point on the DNO's supply transformer is connected to earth?
4. How far from an earth rod does its resistance area extend?
5. How is a ground level voltage gradient for an earth rod avoided?
6. What is created by bonding?
7. List the three common DNO earthing systems.
8. In the formula $Z_s = Z_e + R1 + R2$, what is represented by each of the symbols?
9. What is the maximum disconnection time for a 20A final circuit supplied from a TN system?
10. Which circuits in a caravan should be protected by a 30mA RCD?
11. Why won't a line-neutral fault cause an RCD to trip?
12. How often should an RCD's test facility be operated?

Answers

1. Zero volts
2. Rod electrodes
3. The Neutral
4. Approximately 2.5 to 3 metres
5. By placing underground in a pit
6. An equipotential zone
7. TT, TN-S, TN-C-S
8. Z_s = total loop impedance; Z_e = external loop impedance; R1 = resistance of line conductor; R2 = resistance of protective conductor
9. 0.4 secs
10. All circuits
11. The RCD will see it as a load and not an out of balance fault current
12. Quarterly

CHAPTER 3

Protection

☞ Relevant IET parts, chapters and sections: Part 4, Chapters 41, 42, 43, 44; Part 5, Chapter 53

Important terms/topics covered in this chapter:

- **Arm's reach** A zone of accessibility to touch, extending from any point on a surface where persons usually stand or move about, to the limits which a person can reach with his hand in any direction without assistance
- **Barrier** A part providing a defined degree of protection against contact with live parts, from any usual direction
- **Basic protection** Protection against electric shock under fault-free conditions
- **Circuit protective conductor** A protective conductor connecting exposed conductive parts of equipment to the main earthing terminal
- **Class II equipment** Equipment in which protection against electric shock does not rely on basic insulation only, but in which additional safety precautions such as supplementary insulation are provided. There is no provision for the connection of exposed metalwork of the equipment to a protective conductor, and no reliance upon precautions to be taken in the fixed wiring of the installation
- **Design current** The magnitude of the current intended to be carried by a circuit in normal service
- **Enclosure** A part providing an appropriate degree of protection of equipment against certain external influences and a defined degree of protection against contact with live parts from any direction
- **Exposed conductive part** A conductive part of equipment which can be touched and which is not a live part but which may become live under fault conditions
- **External influence** Any influence external to an electrical installation which affects the design and safe operation of that installation
- **Extraneous conductive part** A conductive part liable to introduce a potential, generally earth potential, and not forming part of the electrical installation

18th Edition IET Wiring Regulations: Explained and Illustrated. 978-1-138-60606-7.

it will also protect against fault currents. However, if there is any doubt the formula should be used.

For example, in Figure 3.6, if I_n has been selected for overload protection, the questions to be asked are as follows:

1.	Is $I_n \geq I_b$?	Yes
2.	Is $I_n \leq I_z$?	Yes
3.	Is $I_2 \geq 1.45 I_z$?	Yes

Then, if the device has a rated breaking capacity not less than the PFC, it can be considered to give protection against fault current also.

When an installation is being designed, the PFC at every relevant point must be determined, by either calculation or measurement. The value will decrease as we move farther away from the intake position (resistance increases with length). Thus, if the breaking capacity of the lowest rated fuse in the installation is greater than the PFC at the origin of the supply, there is no need to determine the value except at the origin.

Discrimination (IET Regulation 536.2)

When we discriminate, we indicate our preference over other choices: this house rather than that house, for example. With protection we have to ensure that the correct device operates when there is a fault. Hence, a 13 A BS 1362 plug fuse should operate before the main circuit fuse. Logically, protection starts at the origin of an installation with a large device and progresses down the chain with smaller and smaller sizes.

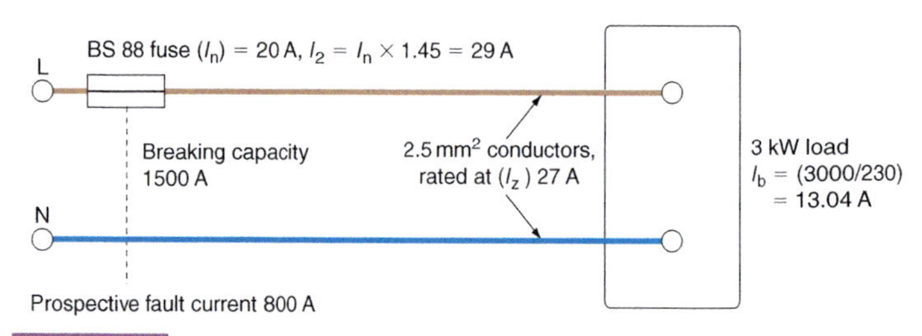

FIGURE 3.6 Breaking capacity and PFC.

Simply because protective devices have different ratings, it cannot be assumed that discrimination is achieved. This is especially the case where a mixture of different types of device is used. However, as a general rule a 2:1 ratio with the lower-rated devices will be satisfactory. The table on page 57 shows how fuse links may be chosen to ensure discrimination.

Fuses will give discrimination if the figure in column 3 does not exceed the figure in column 2. Hence:

a 2 A fuse will discriminate with a 4 A fuse
a 4 A fuse will discriminate with a 6 A fuse
a 6 A fuse will *not* discriminate with a 10 A fuse
a 10 A fuse will discriminate with a 16 A fuse.

All other fuses will *not* discriminate with the next highest fuse, and in some cases several sizes higher are needed, e.g. a 250 A fuse will only discriminate with a 400 A fuse.

Position of protective devices (IET Regulations 433.2 and 434.2)

When there is a reduction in the current-carrying capacity of a conductor, a protective device is required. There are, however, some exceptions to this requirement; these are listed quite clearly in Sections 433 and 434 of the IET Regulations. As an example, protection is not needed in a ceiling rose where the cable size changes from 1.0 mm^2 to, say, 0.5 mm^2 for the lampholder flex. This is permitted as it is not expected that lamps will cause overloads.

PROTECTION AGAINST OVERVOLTAGE (IET REGULATIONS SECTION 443)

This chapter deals with the requirements of an electrical installation to withstand overvoltages caused by lightning or switching surges. Provided equipment, by virtue of its product specification, can withstand at least the values give in the tables, then no action is needed. However IT equipment is very susceptible and unless integral provision exists, surge protective devices may need to be installed.

PROTECTION AGAINST UNDERVOLTAGE (IET REGULATIONS SECTION 445)

From the point of view of danger in the event of a drop or loss of voltage, the protection should prevent automatic restarting of machinery, etc. In fact, such protection is an integral part of motor starters in the form of the control circuit.

Questions

1. What would be the minimum IP code for an item of equipment subject to splashes?
2. What IK code would indicate protection against an impact of 5 joules?
3. What is 'basic protection'?
4. What is 'fault protection'?
5. What system is recommended for the supply to portable tools on a construction site?
6. What is the fusing factor of a BS88 fuse?
7. What is the relationship between the breaking capacity of a protective device and prospective fault current?
8. What is total let-through energy?
9. What is the relationship between 'let-through' energy and the thermal capacity of a cable?
10. What is meant by discrimination between protective devices?

Answers

1. IP X4
2. IK 08
3. Protection against shock under fault free conditions
4. Protection against shock under single fault conditions
5. Reduced low voltage
6. 1.45
7. It must break, and in the case of a circuit breaker, make, the PFC at the point that it is installed without damage to surrounding equipment etc.
8. The electrical energy that a protective device lets through for the duration of a fault, and is given by I^2t
9. The let-through energy must not exceed the thermal capacity of the cable hence $I^2t \leq K^2S^2$
10. Ensuring that a minor protective device operates before a major device

CHAPTER 4

Isolation Switching and Control

Important terms/topics covered in this chapter:

- **Emergency switching** Rapid cutting off of electrical energy to remove any hazard to persons, livestock or property which may occur unexpectedly
- **Isolation** Cutting off an electrical installation, a circuit or an item of equipment from every source of electrical energy
- **Mechanical maintenance** The replacement, refurbishment or cleaning of lamps and non-electrical parts of equipment, plant and machinery
- **Switch** A mechanical switching device capable of making, carrying and breaking current under normal circuit conditions, which may include specified overload conditions, and also of carrying, for a specified time, currents under specified abnormal conditions such as those of short circuit

By the end of this chapter the reader should:

- be aware of the reasons for isolation and switching,
- know the types of switching required in various types of installation,
- be able to select various general types of device to perform relevant switching activities.

ISOLATION AND SWITCHING (IET REGULATIONS CHAPTERS 46 AND 53)

All installations, whether they be the whole or part, must have a means of isolation and switching for various reasons. These are:

1. To remove possible dangers associated with the installation/operation/testing of electrical installations.
2. To provide a means of functional switching and control.

18th Edition IET Wiring Regulations: Explained and Illustrated. 978-1-138-60606-7.

As we know that the maximum voltage drop in this instance (230 V) is 11.5 V, we can determine the maximum length by transposing the formula:

$$\text{Maximum length} = \frac{V_c \times 1000}{\text{mV} \times I_b} = \frac{11.5 \times 1000}{2.8 \times 23} = 178\ \text{m}$$

There are other constraints, however, which may not permit such a length.

SHOCK RISK (IET REGULATIONS SECTION 411)

This topic has already been discussed in full in Chapter 2. To recap, however, the actual loop impedance Z_s should not exceed those values given in Tables 41.2, 41.3 and 41.4 of the IET Regulations. This ensures that circuits feeding final and distribution circuits will be disconnected, in the event of an earth fault, in the required time.

Remember $Z_s = Z_e + R_1 + R_2$.

THERMAL CONSTRAINTS (IET REGULATIONS SECTION 543)

The IET Regulations require that we either select or check the size of a cpc against Table 54.7 of the Regulations, or calculate its size using an adiabatic equation.

Selection of cpc using Table 54.7

Table 54.7 of the Regulations simply tells us that:

1. For line conductors up to and including 16 mm², the cpc should be at least the same size.
2. For sizes between 16 mm² and 35 mm², the cpc should be at least 16 mm².
3. For sizes of line conductor over 35 mm², the cpc should be at least half this size.

This is all very well, but for large sizes of line conductor the cpc is also large and hence costly to supply and install. Also, composite cables such as the typical twin with cpc 6242Y type have cpcs smaller than the line conductor and hence do not comply with Table 54.7.

Calculation of cpc using an adiabatic equation

The adiabatic equation

$$S = \frac{\sqrt{I^2 t}}{k}$$

enables us to check on a selected size of cable, or on an actual size in a multicore cable. In order to apply the equation we need first to calculate the earth fault current from:

$$I = \frac{U_0}{Z_s}$$

where U_0 is the nominal line voltage to earth (usually 230 V) and Z_s is the actual earth fault loop impedance. Next we select a k factor from Tables 54.2 to 54.7 of the Regulations, and then determine the disconnection time t from the relevant curve.

For those unfamiliar with such curves, using them may appear a daunting task. A brief explanation may help to dispel any fears. Referring to any of the curves for fuses in Appendix 3 of the IET Regulations, we can see that the current scale goes from 1 A to 10 000 A, and the time scale from 0.01 s to 10 000 s. One can imagine the difficulty in drawing a scale between 1 A and 10 000 A in divisions of 1 A, and so a logarithmic scale is used. This cramps the large scale into a small area. All the subdivisions between the major divisions increase in equal amounts depending on the major division boundaries; for example, all the subdivisions between 100 and 1000 are in amounts of 100 (Figure 5.7).

Figures 5.8 and 5.9 give the IET Regulations time/current curves for BS 88 fuses. Referring to the appropriate curve for a 32 A fuse (Figure 5.9), we find that a fault current of 200 A will cause disconnection of the supply in 0.6 s.

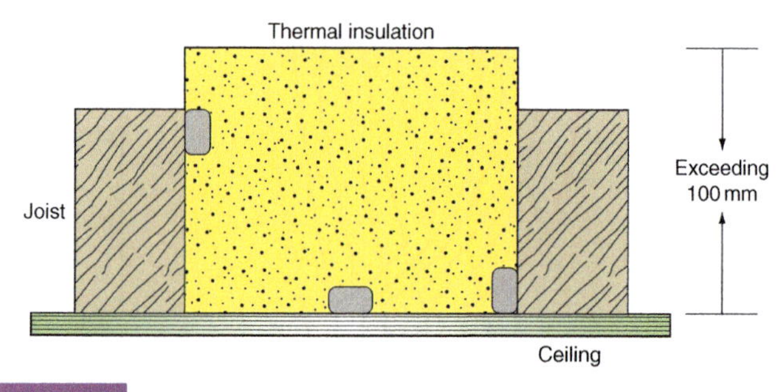

FIGURE 5.13 Method 101.

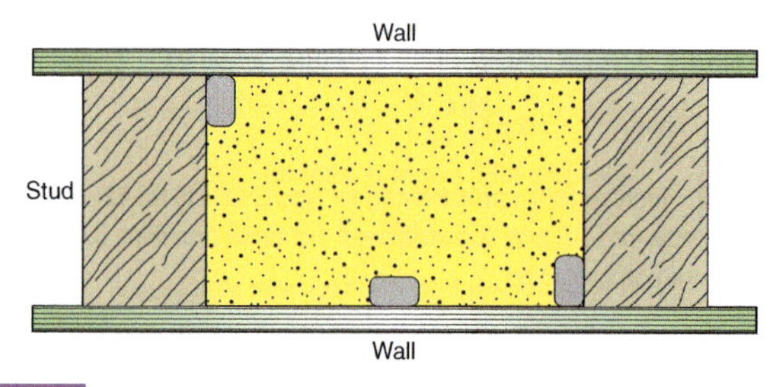

FIGURE 5.14 Method 102.

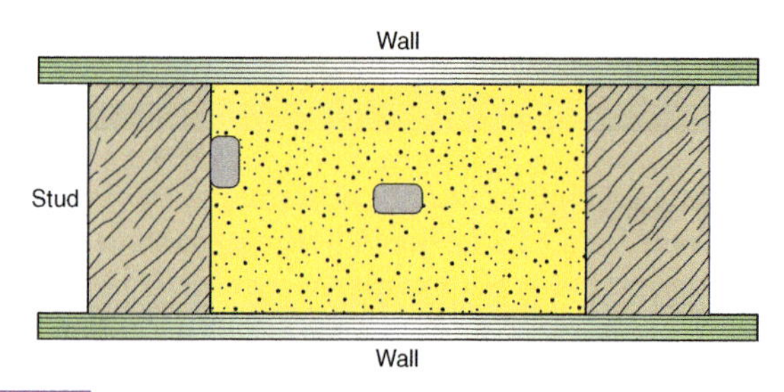

FIGURE 5.15 Method 103.

The mains intake position is at high level and comprises an 80 A BS 1361 230 V main fuse, an 80 A rated meter and a six-way 80 A consumer unit housing BS EN 60898 Type B CBs as follows:

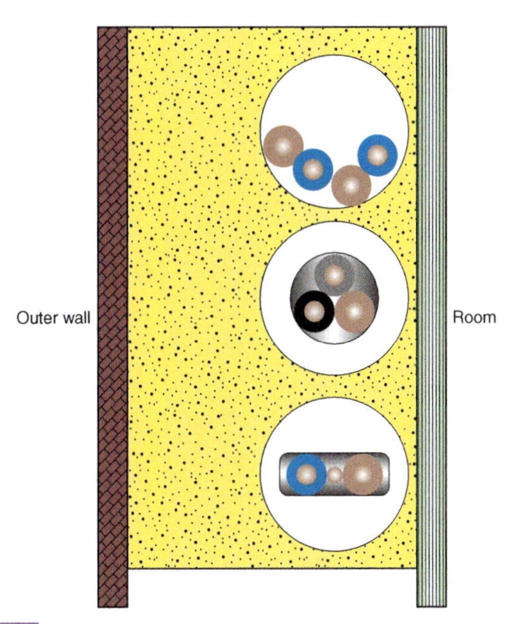

FIGURE 5.16 Method A.

Ring circuit	32 A
Lighting circuit	6 A
Immersion heater circuit	16 A
Cooker circuit	32 A
Shower circuit	32 A
Spare way	–

The cooker is rated at 30 A, with no socket in the cooker unit. The main tails are 16.0 mm^2 double-insulated PVC, with a 6.0 mm^2 earthing conductor. There is no main protective bonding. The earthing system is TN-S, with an external loop impedance Z_e of 0.3 ohms. The prospective short-circuit current (PSCC) at the origin has been measured as 800 A. The roof space is insulated to full depth of the ceiling

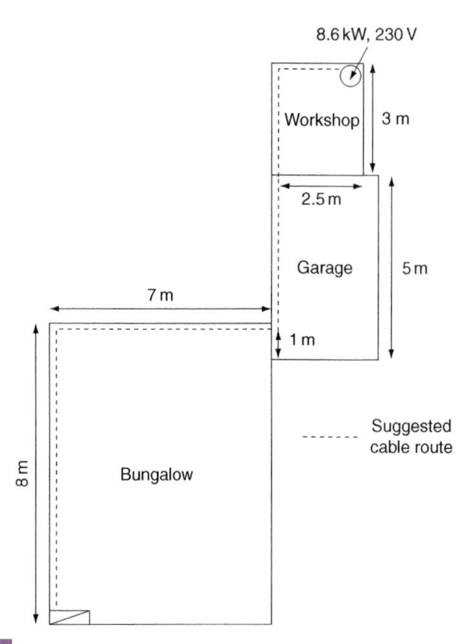

FIGURE 5.19 Diagram for example on page 79.

Reference to the current rating tables in the IET Regulations will show that the existing main tails are too small and should be up-rated. So, the addition of another 8.6 kW of load is not possible with the present arrangement.

The current taken by the kiln is 8600/230 = 37.4 A. Therefore, the new maximum demand is 100 + 37.4 = 137.4 A.

Supply details are: single-phase 230 V, 50 Hz earthing; TN-S PSCC at origin (measured): 800 A.

Decisions must now be made as to the type of cable, the installation method and the type of protective device. As the existing arrangement is not satisfactory, the supply authority must be informed of the new maximum demand, as a larger main fuse and service cable may be required.

SIZING THE MAIN TAILS

1. The new load on the existing consumer unit will be 137.4 A. From the IET Regulations, the cable size is 25.0 mm^2.
2. The earthing conductor size, from the IET Regulations, will be 16.0 mm^2. The main protective bonding conductor size, from the IET Regulations, will be 10.0 mm^2.

For a domestic installation such as this, a PVC flat twin cable, clipped direct (avoiding any thermal insulation) through the loft space and the garage, etc., would be most appropriate.

SIZING THE KILN CIRCUIT CABLE

Design current I_b is:

$$I_b = \frac{P}{V} = \frac{8600}{230} = 37.4 \text{ A}$$

Rating and type of protection I_n

As we have seen, the requirement for the rating I_n is that $I_n \geq I_b$. Therefore, using the tables in the IET Regulations, I_n will be 40 A.

Rating factors:

C_a: 0.94
C_g: not applicable
C_f: 0.725 **only** if the fuse is BS 3036 (not applicable here)
C_i: 0.5 if the cable is totally surrounded in thermal insulation (not applicable here).

Tabulated current-carrying capacity of cable

$$I_t = \frac{I_n}{C_a} = \frac{40}{0.94} = 42.5 \text{ A}$$

Cable size based on tabulated current-carrying capacity

Table 4D5A IET Regulations gives a size of 6.0 mm^2 for this value of I_t (method C).

Check on voltage drop

The actual voltage drop is given by:

$$\frac{\text{mV} \times I_b \times 1}{1000} = \frac{7.3 \times 37.4 \times 24.5}{1000} = 6.7 \text{ V (well below the maximum of 11.5 V)}$$

This voltage drop, whilst not causing the kiln to work unsafely, may mean inefficiency, and it is perhaps better to use a 10.0 mm^2 cable.

For a 10.0 mm^2 cable, the voltage drop is checked as:

$$\frac{4.4 \times 37.4 \times 24.5}{1000} = 4.04 \text{ V}$$

Shock risk

The cpc associated with a 10.0 mm^2 twin 6242 Y cable is 4.0 mm^2. Hence, the total loop impedance will be:

$$Z_s = Z_e + \frac{(R_1 + R_2) \times L \times 1.2}{1000} = 0.3 + \frac{6.44 \times 24.5 \times 1.2}{1000} = 0.489\,\Omega$$

Note

6.44 is the tabulated $(R_1 + R_2)$ value and the multiplier 1.2 takes account of the conductor resistance at its operating temperature.

It is likely that the chosen CB will be a type B.

Thermal constraints

We still need to check that the 4.0 mm^2 cpc is large enough to withstand damage under earth fault conditions. So, the fault current would be:

$$I = \frac{U_0}{Z_s} = \frac{230}{0.489} = 470 \text{ A}$$

The disconnection time t for this current for this type of protection (from the relevant curve in the IET Regulations) is as follows.

40 A CB type B = 0.1 s (the actual time is less than this but 0.1 s is the instantaneous time).

From the regulations, the factor for $k = 115$. We can now apply the adiabatic equation:

$$S = \frac{\sqrt{I^2 \times t}}{k} = \frac{\sqrt{470^2 \times 0.1}}{115} = 1.29\ \text{mm}^2$$

Hence, our 4.0 mm² cpc is of adequate size.

Summary

The kiln circuit would be protected by a 40 A BS EN 60898 type B CB and supplied from a spare way in the consumer unit. The main fuse would need to be up-rated to 100 A. The main tails would be changed to 25.0 mm². The earthing conductor would be changed to 16.0 mm².

Main protective bonding conductors would need to be 10.0 mm² twin with earth PVC cable.

Questions

1. What should be carried out prior to the commencement of design calculations?
2. What would be the design current of a 30 kW 400 V three-phase motor that is 85% efficient and has a p.f of 0.75?
3. What size of BS EN 60898 Type B circuit breaker would be required for a circuit with a design current of 68 A?
4. What is the rating factor for a group of 5 identical multicore cables installed, bunched in air to method C?
5. What de-rating factor would be applied to a cable passing horizontally through a thermally insulated stud wall 100 mm thick?
6. Determine the tabulated current carrying capacity I_t of a cable protected by a 15A BS 3036 fuse. The cable is not subjected to any adverse effects along its run.
7. Calculate the voltage drop for a 25 m length of 4.0 mm² flat twin with cpc cable if the protective device is rated at 32 A. Is this volt drop acceptable?
8. Is there a shock risk if a 20-metre long final circuit is protected by a 32 A BS 88-2 fuse, the conductor is 70° thermoplastic 6.0 mm2 with a 2.5 mm² cpc and the external loop impedance is 0.3Ω?
9. What current would flow in the event of an earth fault at the end of the circuit in question 8 above?
10. What minimum size of cpc would be required for a circuit if a fault current of 500 A disconnected in 0.2 seconds in a cable having a 'k' factor of 100?

Answers

1. An assessment of general characteristics
2. 68 A (30,000 × 100/√3 × 400 × 85 × 0.75)
3. 80 A (table 41.3)
4. 0.6 (table 4C1)
5. 0.78 (table 52.2)
6. 20.7 A (15/0.725)
7. 8.8 V (11 × 32 × 25/1000). Yes, acceptable as max. is 11.5 V
8. No, actual is 0.55 Ω (0.3 + (10.49 × 20 × 1.2)/1000). Max. is 1.3 Ω
9. 418 A ($I = U_0/Z_s$ = 230/0.55
10. 2.5mm² ($s = \sqrt{I^2t}/k = \sqrt{500^2 \times 0.2}/100$)

CHAPTER 6

Inspection and Testing

Important terms/topics covered in this chapter:

- **Earth electrode** A conductor or group of conductors in intimate contact with and providing an electrical connection with earth
- **Earth fault loop impedance** The impedance of the earth fault loop (line-to-earth loop) starting and ending at the point of earth fault
- **Residual current device (RCD)** An electromechanical switching device or association of devices intended to cause the opening of the contacts when the residual current attains a given value under specified conditions
- **Ring final circuit** A final circuit arranged in the form of a ring and connected to a single point of supply

By the end of this chapter the reader should:

- know the sequence of tests to be carried out on an installation,
- be aware of the need to zero or null instruments before conducting low resistance tests,
- know the general procedure for conducting each of the tests,
- be able to interpret test results.

TESTING SEQUENCE (PART 6)

Having designed our installation, selected the appropriate materials and equipment, and installed the system, it now remains to put it into service. However, before this happens, the installation must be tested and inspected to ensure that it complies, as far as is practicable, with the IET Regulations. Note the word 'practicable'; it would be unreasonable, for example, to expect the whole length of a circuit cable to be inspected for defects, as this may mean lifting floorboards, etc.

Part 6 of the IET Regulations gives details of testing and inspection requirements. Unfortunately, these requirements pre-suppose that the

18th Edition IET Wiring Regulations: Explained and Illustrated. 978-1-138-60606-7.

How, then, do we conduct a test to establish continuity of main or supplementary bonding conductors? Quite simple really: just connect the leads from the continuity tester to the ends of the bonding conductor (Figure 6.1). One end should be disconnected from its bonding clamp, otherwise any measurement may include the resistance of parallel paths of other earthed metalwork. Remember to zero or null the instrument first or, if this facility is not available, record the resistance of the test leads so that this value can be subtracted from the test reading.

Important Note

If the installation is in operation, then never disconnect main bonding conductors unless the supply can be isolated. Without isolation, persons and livestock are at risk of electric shock.

The continuity of circuit protective conductors may be established in the same way, but a second method is preferred, as the results of this second test indicate the value of $(R_1 + R_2)$ for the circuit in question.

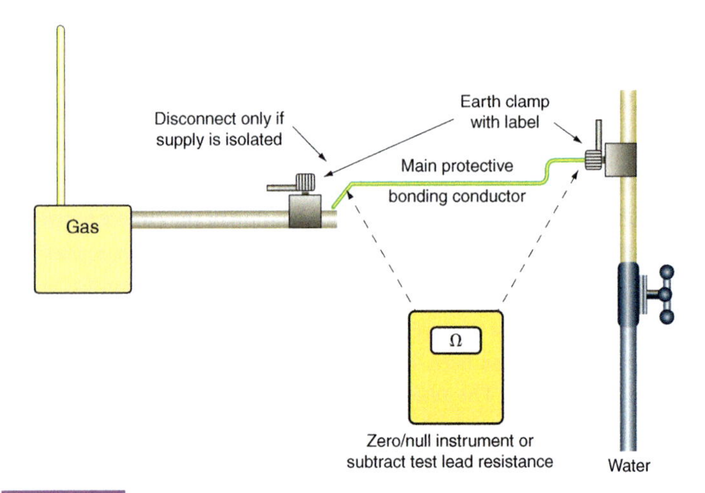

FIGURE 6.1 Continuity of main protective bonding conductor.

The test is conducted in the following manner:

1. Temporarily link together the line conductor and cpc of the circuit concerned in the distribution board or consumer unit.
2. Test between line and cpc at each outlet in the circuit. A reading indicates continuity.
3. Record the test result obtained at the furthest point in the circuit. This value is $(R_1 + R_2)$ for the circuit.

Figure 6.2 illustrates the above method.

There may be some difficulty in determining the $(R_1 + R_2)$ values of circuits in installations that comprise steel conduit and trunking, and/or SWA and mims cables because of the parallel earth paths that are likely to exist. In these cases, continuity tests may have to be carried out at the installation stage before accessories are connected or terminations made off as well as after completion.

Continuity of ring final circuit conductors

There are two main reasons for conducting this test:

1. To establish that interconnections in the ring do not exist.
2. To ensure that the circuit conductors are continuous, and indicate the value of $(R_1 + R_2)$ for the ring.

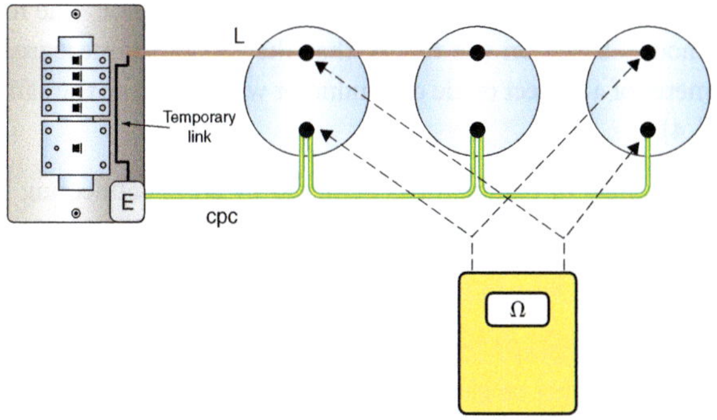

FIGURE 6.2 cpc continuity.

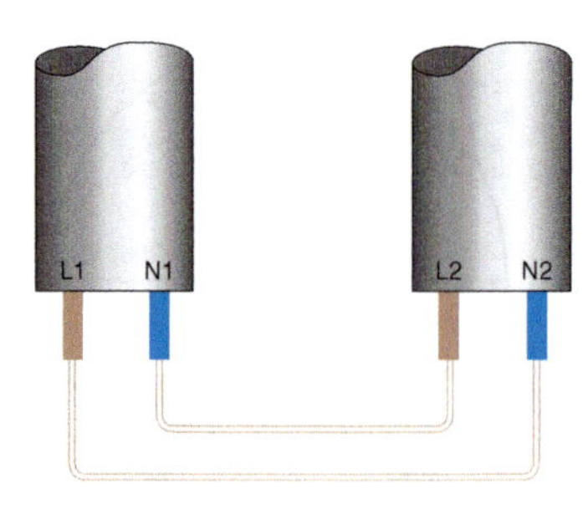

FIGURE 6.6 L and N cross-connection.

midpoint of the ring, then decreasing values back towards the interconnection. If a break had occurred at point Y then the readings from socket S would increase to a maximum at socket P. One or two high readings are likely to indicate either loose connections or spurs. A null reading, i.e. an open circuit indication, is probably a reverse polarity, either line–cpc or neutral–cpc reversal. These faults would clearly be rectified and the test at the suspect socket(s) repeated.

5. *Repeat the above procedure, but in this case cross-connect the line and cpc loops*. In this instance, if the cable is of the flat twin type, the readings at each socket will very slightly increase and then decrease around the ring. This difference, due to the line and cpc being different sizes, will not be significant enough to cause any concern. The measured value is very important it is $(R_1 + R_2)$ for the ring.

As before, loose connections, spurs and, in this case, L–N cross-polarity will be picked up.

The details that follow are typical approximate ohmic values for a healthy 70 m ring final circuit wired in 2.5 mm²/1.5 mm² flat twin and cpc cable:

Initial Measurements	L1–L2	N1–N2	cpc1–cpc2
Reading at each socket	0.26 Ω	0.26 Ω	between 0.32 Ω and 0.34 Ω
For spurs, each metre in length will add the following resistance to the above values	0.015 Ω	0.015 Ω	0.02 Ω

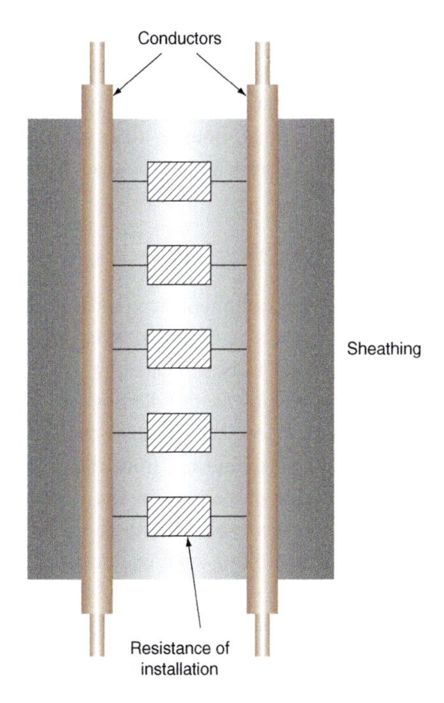

FIGURE 6.7 Cable insulation resistance.

Insulation resistance

This is probably the most used and yet most abused test of them all. Affectionately known as 'meggering', an insulation resistance test is performed in order to ensure that the insulation of conductors, accessories and equipment is in a healthy condition, and will prevent dangerous leakage currents between conductors and between conductors and earth. It also indicates whether any short circuits exist.

Insulation resistance is the resistance measured between conductors and is made up of countless millions of resistances in parallel (Figure 6.7).

The more resistances there are in parallel, the lower the overall resistance, and in consequence, the longer a cable the lower the insulation resistance. Add to this the fact that almost all installation circuits are also wired in parallel, and it becomes apparent that tests on large installations may give, if measured as a whole, pessimistically low values, even if there are no faults. Under these circumstances, it is usual to break down such large installations into smaller sections, floor by floor, distribution circuit by distribution circuit, etc. This also helps, in the case of periodic testing, to minimize disruption.

The test procedure, then, is as follows:

1. Disconnect all items of equipment such as capacitors and indicator lamps as these are likely to give misleading results. Remove any items of equipment likely to be damaged by the test, such as dimmer switches, electronic timers, etc. Remove all lamps and accessories and disconnect fluorescent and discharge fittings. Ensure that the installation is disconnected from the supply, all fuses are in place, and CBs and switches are in the on position. In some instances it may be impracticable to remove lamps, etc. and in this case the local switch controlling such equipment may be left in the off position.
2. Join together all live conductors of the supply and test between this join and earth. Alternatively, test between each live conductor and earth in turn.
3. Test between line and neutral. For three-phase systems, join together all lines and test between this join and neutral. Then test between each of the lines. Alternatively, test between each of the live conductors in turn. Installations incorporating two-way lighting systems should be tested twice with the two-way switches in alternative positions.

Table 6.2 gives the test voltages and minimum values of insulation resistance for ELV and LV systems.

Table 6.2 Insulation Resistance Test Values

System	Test Voltage	Minimum Insulation Resistance
SELV and PELV	250 V d.c.	0.5 MΩ
LV up to 500 V	500 V d.c.	1.0 MΩ
Over 500 V	1000 V d.c.	1.0 MΩ

If a value of less than 2 MΩ is recorded it may indicate a situation where a fault is developing, but as yet still complies with the minimum permissible value. In this case, each circuit should be tested separately to identify any that are suspect.

Polarity

This simple test, often overlooked, is just as important as all the others, and many serious injuries and electrocutions could have been prevented if only polarity checks had been carried out.

The requirements are:

- all fuses and single pole switches are in the line conductor
- the centre contact of an Edison screw type lampholder is connected to the line conductor
- all socket outlets and similar accessories are correctly wired.

Although polarity is towards the end of the recommended test sequence, it would seem sensible, on lighting circuits for example, to conduct this test at the same time as that for continuity of cpcs (Figure 6.8).

As discussed earlier, polarity on ring final circuit conductors is achieved simply by conducting the ring circuit test. For radial socket outlet circuits, however, this is a little more difficult. The continuity of the cpc will have already been proved by linking line and cpc and measuring between the same terminals at each socket. Whilst a

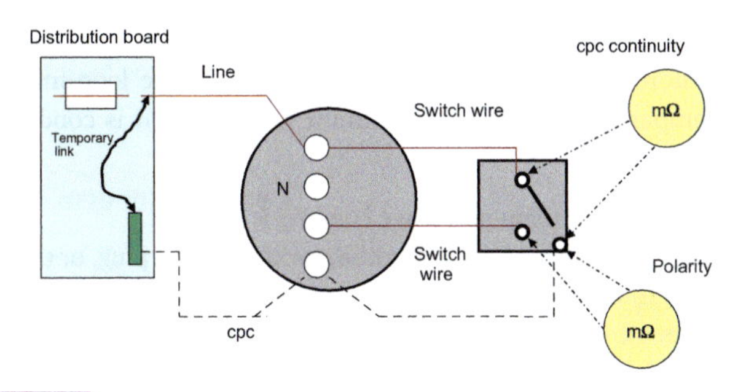

FIGURE 6.8 Lighting polarity.

Table 6.3 Corrected Maximum Z_s Values for Comparison with Measured Values

PROTECTIVE DEVICE		PROTECTION RATING																		
		5	**6**	**10**	**15**	**16**	**20**	**25**	**30**	**32**	**40**	**45**	**50**	**60**	**63**	**80**	**100**	**125**	**160**	**200**
BS EN 60898 and 61009 type **B**	0.4s & 5s		*5.82*	*3.5*		*2.19*	*1.75*	*1.4*		*1.09*	*0.87*		*0.7*		*0.56*	*0.5*	*0.35*	*0.29*		
BS EN 60898 and 61009 type **C**	0.4s & 5s		*2.9*	*1.75*		*1.09*	*0.87*	*0.71*		*0.54*	*0.44*		*0.35*		*0.28*	*0.2*	*0.17*	*0.14*		
BS EN 60898 and 61009 type **D**	0.4s		*1.46*	*0.87*		*0.54*	*0.44*	*0.35*		*0.28*	*0.22*		*0.17*		*0.13*	*0.1*	*0.08*	*0.07*		
	5s		*2.91*	*1.75*		*1.09*	*0.87*	*0.69*		*0.55*	*0.44*		*0.35*		*0.27*	*0.22*	*0.17*	*0.14*		
BS 3036 Semi enclosed	0.4s	*7.22*			*1.94*		*1.34*		*0.83*			*0.45*		*0.32*						
	5s	*13.45*			*4.1*		*2.9*		*2*			*1.2*		*0.85*			*0.4*			
		3A*			**13A***															
BS 1362 Cartridge	0.4s	*12.5*			*1.8*															
	5s	*17.63*			*2.9*															

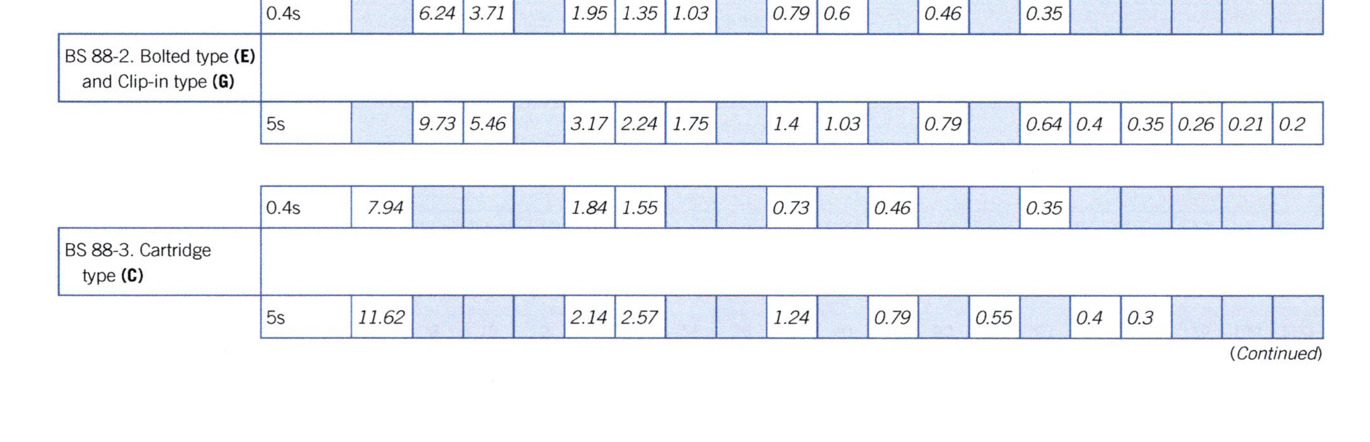

BS 88-2. Bolted type **(E)** and Clip-in type **(G)**	0.4s		*6.24*	*3.71*		*1.95*	*1.35*	*1.03*		*0.79*	*0.6*		*0.46*		*0.35*					
	5s		*9.73*	*5.46*		*3.17*	*2.24*	*1.75*		*1.4*	*1.03*		*0.79*		*0.64*	*0.4*	*0.35*	*0.26*	*0.21*	*0.2*
BS 88-3. Cartridge type **(C)**	0.4s	*7.94*				*1.84*	*1.55*			*0.73*		*0.46*			*0.35*					
	5s	*11.62*				*2.14*	*2.57*			*1.24*		*0.79*		*0.55*		*0.4*	*0.3*			

(*Continued*)

Table 6.3 Corrected Maximum Z_s Values for Comparison with Measured Values – Cont'd

OBSOLETE FUSES & CIRCUIT BREAKERS

MAXIMUM CORRECTED Z_s VALUES		PROTECTION RATING																		
		5	6	10	15	16	20	25	30	32	40	45	50	60	63	80	100	125	160	200
BS 88 – 2.2 & 88-6	0.4s		6.48	3.89		2.05	1.35	1.09		0.79										
	5s		10.26	5.64		3.16	2.2	1.75		1.4	1.03		0.79		0.64	0.5	0.32	0.25	0.19	0.2
BS 1361 Cartridge	0.4s	7.94			2.45		1.29		0.87											
	5s	12.46			3.8		2.13		1.4			0.75		0.53		0.4	0.28			
BS 3871 type 1	0.4s & 5s	8.74	7.22	4.37	2.9	2.73	2.19	1.75	1.45	1.37	1.09	0.97	0.87		0.69					
BS 3871 type 2	0.4s & 5s	5	4.15	2.49	1.66	1.56	1.24	1	0.83	0.78	0.64	0.55	0.49		0.42					
BS 3871 type 3	0.4s & 5s	3.5	2.85	1.75	1.16	1.09	0.87	0.7	0.58	0.54	0.44	0.39	0.35		0.28					

Table 6.4 RCD Tripping Times

RCD Type	Half Rated	Full Trip Current
BS 4293 sockets	No trip	Less than 200 ms
BS 4293 with time delay	No trip	½ time delay 1200 ms
BS EN 61009 or BS EN 61009 RCBO	No trip	300 ms
As above, but type S with time delay	No trip	130–500 ms

Additional protection RCD/RCBO operation

Where RCDs/RCBOs are fitted, it is essential that they operate within set parameters. The RCD testers used are designed to do just this, and the basic tests required are as follows:

1. Set the test instrument to the rating of the RCD.
2. Set the test instrument to half-rated trip.
3. Operate the instrument and the RCD should not trip.
4. Set the instrument to deliver the full rated tripping current of the RCD, $I_{\Delta n}$.
5. Operate the instrument and the RCD should trip out in the required time.
6. For RCDs rated at 30 mA or less set the instrument to deliver 5 times the rated tripping current of the RCD, $5I_{\Delta n}$.
7. Operate the instrument and the RCD should trip out in 40 ms.

Table 6.4 gives further details.

Prospective fault current

Prospective fault current (PFC) has to be determined at the origin of the installation. This is achieved by enquiry, calculation or measurement.

Note

For domestic premises with consumer units to BS EN 61439-3 and a Distribution Network Operator declared PFC of 16 ka, There is no requirement to measure or calculate the PFC.

Check of phase sequence

Where multi-phase systems are used there is a high possibility that phase sequence will need to be checked.

This is done with the use of a phase rotation indicator, which, simplistically, is a small three-phase motor.

Functional testing

All RCDs have a built-in test facility in the form of a test button. Operating this test facility creates an artificial out-of-balance condition that causes the device to trip. This only checks the mechanics of the tripping operation; it is not a substitute for the tests just discussed.

All other items of equipment such as switchgear, control gear, interlocks, etc. must be checked to ensure that they are correctly mounted and adjusted and that they function correctly.

Verification of voltage drop

Where required the voltage drop to the furthest point of a circuit should be determined. This is not usually needed for initial verification.

Questions

1. Why should a test for continuity of a protective bonding conductor require the disconnection of that conductor?
2. What precaution should be taken before conducting a continuity test on a protective bonding conductor?
3. What is the significance of the reading obtained at the end of a circuit when conducting a cpc continuity test?
4. Apart from identifying interconnections and establishing a value for ($R_1 + R_2$), what other test is automatically conducted when carrying out a ring final circuit continuity test?
5. For an insulation resistance test on a 600 V discharge lighting circuit, state the required test voltage and minimum value of insulation resistance.
6. What instrument is used for conducting a polarity test?
7. Why is a 0.8 factor used when comparing measured values of Z_s with maximum tabulated values?
8. Why must the earthing conductor of an installation be disconnected when conducting a test for external loop impedance?
9. What are the test requirements for a 20 mA RCD used for additional protection?
10. What is the purpose of the test function on an RCD?

Answers

1. To avoid parallel paths
2. Isolate the supply
3. $(R_1 + R_2)$
4. Polarity
5. 1000V, 1 MΩ
6. Low resistance ohmmeter
7. to compensate for ambient and conductor temperature
8. To avoid parallel paths
9. Must disconnect in 40ms at $5 \times I_{\Delta n}$ (100 mA)
10. To confirm that its mechanism is functioning correctly

It does not cover public or private events that form part of entertainment activities, which are the subject of BS 7909.

External influences

None particularly specified. Clearly they must be considered and addressed accordingly.

Wiring

Armoured or mechanically protected cables where there is a risk of mechanical damage.

Cables shall have a minimum conductor size of 1.5 mm^2.

Protection

Against shock:

1. Supply cables to a stand or unit, etc. must be protected at the cable origin by a time-delayed RCD of residual current rating not exceeding 300 mA.
2. All socket outlet circuits not exceeding 32 A and all other final circuits, excepting emergency lighting, shall have additional protection by 30 mA RCDs.
3. Any metallic structural parts accessible from within the unit stand, etc. shall be connected by a main protective bonding conductor to the main earthing terminal of the unit.

Against thermal effects:

1. Clearly in this case all luminaires, spot lights, etc. should be placed in such positions as not to cause a build-up of excessive heat that could result in fire or burns.

Isolation

Every unit, etc. should have a readily accessible and identifiable means of isolation.

Inspection and testing

Tongue in cheek here! Every installation **should** be inspected and tested on site in accordance with Part 6 of BS 7671.

BS 7671 SECTION 712: SOLAR PHOTOVOLTAIC (PV) SUPPLY SYSTEMS

These are basically solar panels generating d.c. which is then converted to a.c. via an invertor. Those dealt with in BS 7671 relate to those systems that are used to 'top up' the normal supply.

There is a need for consideration of the external influences that may affect cabling from the solar units outside to control gear inside.

There must be protection against overcurrent and a provision made for isolation on both the d.c. and a.c. sides of the invertor.

As the systems can be used in parallel with or as a switched alternative to the public supply, reference should be made to Chapter 55 of BS 7671.

BS 7671 SECTION 714: OUTDOOR LIGHTING INSTALLATIONS

This section applies, for example, to lighting installations for, roads, car parks, monuments, road signs, town plans etc. but, interestingly not to a luminaire fixed to the outside wall of a premises and fed from the internal wiring.

The method of protection against electric shock for these installations is the standard Automatic Disconnection of Supply with a maximum disconnection time of 5 s.

For items such as town plans, telephone kiosks, bus shelters etc., 30 mA or less RCD protection is required.

External influences should be addressed and equipment should be at least IP33 rated.

BS 7671 SECTION 722: ELECTRIC VEHICLE CHARGING INSTALLATIONS

There are four methods of electric vehicle (EV) charging known as Modes and are as follows:

Mode 1

Connecting the EV to a standard earthed socket outlet not exceeding 16A e.g. BS 1363 or BS EN 60309.

Mode 2

As Mode 1 but up to 32A and a special cable that has a connector/control box with an extra pin at the EV end. This is known as a 'pilot control function' and ensures that current can only flow if the earthing is intact. There must also be RCD protection between the EV and the control box.

Mode 3

This requires the use of dedicated EV supply equipment (EVSE) which incorporates the pilot control function etc. The connectors/plugs are to have control and signal pins which ensure the EVSE socket cannot be energised if an EV is not connected.

Mode 4

In this instance, the a.c. mains is converted to d.c. via a charger incorporating the pilot control function. The EV is connected to this charging unit.

EV charging point equipment outdoors should be at least IP44 rated.

BS 7671 SECTION 729: OPERATING AND MAINTENANCE GANGWAYS

Such gangways are likely to be found in restricted areas that are typical of switchrooms, etc., where protection against contact with live electrical parts of equipment is provided by barriers or enclosures or obstacles, the latter having to be under the control of a skilled persons.

Main points:

- Restricted areas must be clearly and visibly marked.
- Access to unauthorized persons is not permitted.

BS 7671 SECTION 753: FLOOR AND CEILING HEATING SYSTEMS

Systems referred to in this section are those used for thermal storage heating or direct heating.

Protection

Against shock:

1. Automatic disconnection of supply with disconnection achieved by 30 mA RCD.
2. Additional protection for Class II equipment by 30 mA RCDs.
3. Heating systems provided without exposed conductive parts shall have a metallic grid of spacing not more than 300 mm installed on site above a floor system or below a ceiling system and connected to the protective conductor of the system.

Against thermal effects:

1. Where skin or footwear may come into contact with floors the temperature shall be limited, for example to 30°C.
2. To protect against overheating of these systems the temperature of any zone should be limited to a maximum of 80°C.

External influences

Minimum of IPX1 for ceilings and IPX7 for floors.

The designer must provide a comprehensive and detailed plan of the installation which should be fixed on or adjacent to the system distribution board.

Questions

1. When is the space under a bath or shower outside of all the zones?
2. What is the IP rating for equipment in zone 1 of a swimming pool where jets are use for cleaning purposes?
3. What is the maximum height above floor level of zone 2 of a sauna?
4. What rating of RCD is required for a 63 A socket outlet on a construction site?
5. How deep should a cable be buried in arable ground on a farm?
6. What is the maximum rating of an RCD used in conjunction with Class II fixed equipment in a conductive location with restricted movement?

- For closed restricted areas, doors must allow for easy evacuation without the use of a key or tool.
- Gangways must be wide enough for easy access for working and for emergency evacuation.
- Gangways must permit at least a 90° opening of equipment doors etc.
- Gangways longer than 10 m must be accessible from both ends.

BS 7671 SECTION 730: ONSHORE UNITS OF ELECTRICAL SHORE CONNECTIONS FOR INLAND NAVIGATION VESSELS

This section deals with supplying vessels such as tug boats, transport barges, pleasure boats etc and is very similar to the requirements for marinas. In this instance the supplies are likely to be larger and hence socket outlets up to 63 A need to have RCD protection not exceeding 30 mA, and over 63 A, RCD protection not exceeding 300 mA.

BS 7671 SECTION 740: AMUSEMENT DEVICES, FAIRGROUNDS, CIRCUSES, ETC.

This is not an area that is familiar to most installation electricians and hence will only be dealt with very briefly.

The requirements of this section are very similar to those of Section 711 Exhibitions, shows, etc. and parts of Section 706 Agricultural locations (because of animals) regarding supplementary bonding.

For example, additional protection by 30 mA is required for:

1. Lighting circuits, except those that are placed out of arm's reach and not supplied via socket outlets.
2. All socket outlet circuits rated up to 32 A.
3. Mobile equipment supplied by a flexible cable rated up to 32 A.

Automatic disconnection of supply must be by an RCD.

Equipment should be rated to at least IP44.

The installation between the origin and any equipment should be inspected and tested after each assembly on site.

7. What is the impact code for equipment subject to (AG 3) mechanical stress on a caravan park?
8. What is the minimum height of a socket outlet above the highest water level for a walkway in a marina?
9. What is the room Group reference for an operating theatre in a hospital?
10. What rating and type of RCD should be provided at the origin of the supply cable to a stand in an exhibition?
11. For a PV installation, where should inverter isolation be provided?
12. Which socket outlets associated with mobile units should be 30 mA, or less, protected?
13. What is the maximum height above ground level for the inlet to a caravan?
14. Above what distance should a gangway be accessible from both ends?
15. What is the minimum IP rating for equipment in a fairground?
16. What is the minimum IP rating for ceiling heating systems?
17. What is the maximum disconnection time for outdoor lighting installations?
18. What is the maximum a.c. voltage for an extra low voltage lighting system with bare conductors?

Answers

1. When accessible only by a tool
2. IPX5
3. 1 metre
4. 500 mA
5. 1 metre
6. 30 mA or less
7. IK08
8. 300 mm
9. Group 2
10. 300 mA time delayed or 'S' type
11. On both the a.c. and the d.c. side
12. Those used for equipment outside the unit
13. 1.8 m
14. 10 m
15. IP44
16. IPX1
17. 5 s
18. 25 V

22. Which of the following sources of supply would be suitable for SELV and PELV circuits?
   a. Public supply at 230 V a.c.
   b. Private generator at 110 V a.c.
   c. A safety isolating transformer
   d. A low voltage d.c. supply

23. Which of the following formulae is used where there is doubt regarding the effectiveness of supplementary bonding?
   a. $R \geq 50\ V/I_a$
   b. $R \leq 50\ V/I_n$
   c. $R \leq 50\ V/I_a$
   d. $R \leq 50\ V/Iz$

24. In order to protect against burns, a non-metallic electrical part intended to be touched but not hand held should not attain a surface temperature in excess of
   a. 65 °C
   b. 70 °C
   c. 80 °C
   d. 90 °C

25. The disconnection time for a 16.0 mm^2 conductor having a 'k' factor of 115 and carrying a fault current of 13 kA is
   a. 0.0125 s
   b. 0.02 s
   c. 0.14 s
   d. 2.26 s

26. The minimum impulse withstand voltage for a 230 V electricity meter is
   a. 1.5 kV
   b. 2.5 kV
   c. 4 kV
   d. 6 kV

27. A conductor marked green-and-yellow throughout its length with additional blue markings at the termination is a
   a. line conductor
   b. neutral conductor
   c. PEN conductor
   d. circuit protective conductor

28. The notice relating to an RCD requires the device to be operated via its test button
   a. monthly
   b. quarterly
   c. half yearly
   d. annually

29. One method of protecting a cable, that passes through a ceiling joist from damage, is to install it at a vertical distance from the top or bottom of the joist of at least
   a. 10 mm
   b. 25 mm
   c. 30 mm
   d. 50 mm

30. The derating factor for a cable surrounded by thermal insulation for 200 mm is
   a. 0.5
   b. 0.51
   c. 0.63
   d. 0.78

31. Where underground telecommunication and power cables cross, the minimum clearance to be maintained between them is
   a. 50 mm
   b. 100 mm
   c. 200 mm
   d. 0.5 m

32. The magnetic circuit of an RCD shall enclose
   a. all live conductors
   b. all line conductors
   c. all circuit conductors
   d. PEN conductors

33. Where an RCD is used for protection against fire, it shall have a maximum rating of
   a. 30 mA
   b. 100 mA
   c. 300 mA
   d. 500 mA

34. Which of the following devices may **only** be used for functional switching?
   a. A BS EN 60898 circuit breaker
   b. A 16A plug and socket-outlet
   c. A cooker control unit switch
   d. A device with semi-conductors

35. A firefighter's switch should be mounted above ground level at a height of
   a. 2.75 m with the switch OFF position at the top
   b. 2.75 m with the switch ON position at the top
   c. 2.25 m with the switch OFF position at the top
   d. 2.75 m with the switch OFF position at the bottom

36. The minimum size of a buried copper earthing conductor not protected against mechanical damage or corrosion is
    a. 10 mm^2
    b. 16 mm^2
    c. 25 mm^2
    d. 50 mm^2

37. The minimum size of a protective conductor with a 'k' factor of 143 which disconnects in 0.3 s at a fault current of 800 A is
    a. 2.5 mm^2
    b. 4.0 mm^2
    c. 6.0 mm^2
    d. 10.0 mm^2

38. A distribution board containing circuits with high protective conductor currents shall have the information regarding these circuits positioned so as to be visible to
    a. the Distribution Network Operator
    b. the user of the installation
    c. a person modifying or extending a circuit
    d. an inspector

39. The earthing conductor for an installation supplied from a TN-S system is 16 mm^2. The minimum size of a main protective bonding conductor is
    a. 6 mm^2
    b. 10 mm^2
    c. 16 mm^2
    d. 25 mm^2

40. An installation and its generator are not permanently fixed. In a TN, TT or an IT system an RCD shall be installed with a maximum rating of
    a. 500 mA
    b. 300 mA
    c. 100 mA
    d. 30 mA

41. Control equipment incorporating protection against overload shall be provided for every motor having a rating exceeding
    a. 100 W
    b. 0.37 kW
    c. 37 W
    d. 10 kW

42. The inspection of an installation is NOT made to ensure
    a. compliance with section 511 of BS 7671
    b. that equipment is correctly selected and erected
    c. that the requirements of the ESQCR are met
    d. that it is not visible damaged or defective

43. The minimum value of insulation resistance and the applied test voltage for a 400 V circuit is
    a. 1.0MΩ at 500 V a.c.
    b. 1.0MΩ at 500 V d.c.
    c. 0.5MΩ at 250 V d.c.
    d. 1.0MΩ at 1000 V d.c.

44. Where surge protective devices cannot be disconnected for insulation resistance testing, the minimum value of insulation resistance and the applied test voltage are
    a. 1.0 MΩ at 500 V a.c.
    b. 1.0 MΩ at 500 V d.c.
    c. 0.5 MΩ at 250 V d.c.
    d. 1.0 MΩ at 250 V d.c.

45. The correct connection of socket-outlets and similar accessories is established by
    a. functional testing
    b. polarity testing
    c. insulation resistance testing
    d. RCD testing

46. An installation is to have the following tests carried out:
    1. polarity. 2. cpc continuity 3. earth electrode resistance. 4. insulation resistance.
    The correct sequence for these tests is
    a. 1, 2, 3, 4
    b. 4, 2, 3, 1
    c. 3, 1, 4, 2
    d. 2, 4, 1, 3

47. The verification of voltage drop may be achieved by
    a. use of a voltmeter
    b. measuring the circuit impedance
    c. measuring the prospective fault current
    d. functional testing

U

V

W